Table of Contents

Introduction	3
Research Goals of Many Nations	5
Plans Becoming Reality	7
Knowledge for all Humankind	9
An Unprecedented International Laboratory	15
Multipurpose Facilities: Multipurpose Racks, Freezers, and Glove Boxes	19
Biological Research: Incubators, Growth Chambers, and Centrifuges	25

Human Physiology Research:
     Neuroscience, Cardiovascular, Musculoskeletal
     and Exercise Equipment, Radiation Sensors
	31

Physical Science and Materials Research: Fluid Physics, Crystal Growth, and External Test Beds	41
Earth and Space Science *(External and Internal)*: Radiation, Thermal, Solar, and Geophysics	51
ISS Control Centers	59
To Learn More	61
Acronyms	61

This brochure was developed collaboratively by all of the International Partner ISS Program Scientists: Perry Johnson-Green, Ph.D. (CSA), Martin Zell, Ph.D. (ESA), Mr. Tai Nakamura (JAXA), Julie Robinson, Ph.D. (NASA), George Karabadzhak, Ph.D. (Roscosmos), and Mr. Igor Sorokin (Roscosmos).

Executive Editor, Deborah L. Hahn, Ph.D.
Associate Editor, Tara M. Ruttley, Ph.D.
Designer, Amy Gish

Welcome to ISS

The International Space Station (ISS) is an unprecedented achievement in global human endeavors to conceive, plan, build, operate, and utilize a research platform in space.

As we near completion of the ISS on-orbit assembly, including all Partner laboratories and elements, we now turn to the real multifaceted purpose of the ISS. We will use the ISS as a human-tended laboratory in low-Earth orbit to conduct multidiscipline research and technology development and as an outpost to conduct human exploration.

The ISS is uniquely capable of unraveling the mysteries of life on Earth; from biomedical research, to material sciences, to technology advancement, to research on the effects of long-duration spaceflight on the human body.

The ISS is the first step in exploration, from research and discovery, to international cooperation, to commercial development, and to exploring beyond low-Earth orbit.

We look forward to sharing this booklet outlining our ISS research capabilities and potential as we usher in this new phase of on-orbit research.

Program Managers

Mr. Benoit Marcotte
Director General, Operations
Canadian Space Agency (CSA)

Ms. Simonetta Di Pippo
Director of Human Spaceflight
European Space Agency (ESA)

Mr. Koichi Morimoto
Deputy Director-General, Research and Development Bureau
Ministry of Education, Culture, Sports, Science and Technology (MEXT)
Government of Japan

Mr. William H. Gerstenmaier
Associate Administrator for Space Operations
National Aeronautics and Space Administration (NASA)

Mr. Alexey Borisovich Krasnov
Director, Piloted Space Programs
Federal Space Agency (Roscosmos)

Research Goals of Many Nations

It is the unique blend of unified and diversified goals among the world's space agencies that will lead to improvements in life on Earth for all people of all nations. While the various space agency partners may emphasize different aspects of research to achieve their goals in the use of ISS, they are unified in several important overarching goals.

All of the agencies recognize the importance of leveraging ISS as an education platform to encourage and motivate today's youth to pursue careers in math, science, and engineering: **educating the children of today to be the leaders and space explorers of tomorrow.**

Advancing our knowledge in the areas of human physiology, biology, material and physical sciences, and translating that knowledge to health, socioeconomic, and environmental benefits on Earth is another common goal of the agencies: **returning the knowledge gained in space research for the benefit of society.**

Finally, all the agencies are unified in their goals to apply knowledge gained through ISS research in human physiology, radiation, materials science, engineering, biology, fluid physics, and technology to enable future space exploration missions: **preparing for exploring the moon, Mars, and beyond.**

Research Goals of Many Nations

Plans Becoming a Reality

Almost as soon as the ISS was habitable, it was used to study the impact of microgravity and other space effects on several aspects of our daily lives. ISS astronauts conduct science daily across a wide variety of fields including human life sciences, biological science, human physiology, physical and materials science, and Earth and space science. Over 400 experiments have been conducted on the ISS as part of early utilization, over 9 years of continuous research.

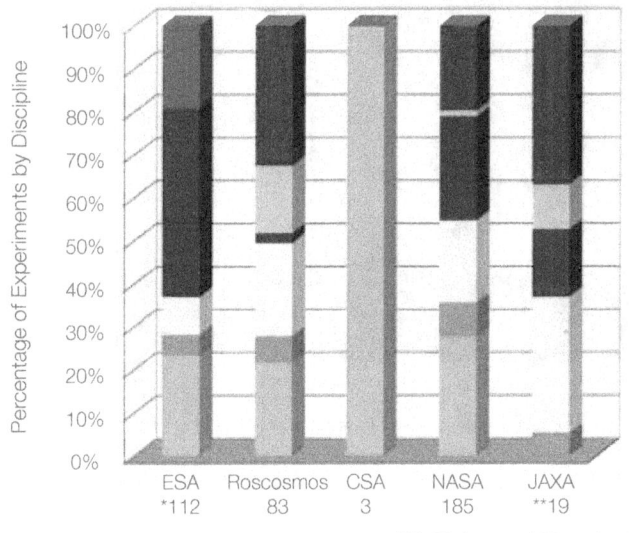

Number of Experiments Performed Through Expedition 18 (April 2009)

- Human Research
- Education
- Technology
- Physical and Materials Science
- Biology and Biotechnology
- Earth and Space Science
- Other

*30 ESA experiments performed since *Columbus* commissioning on 2/17/08
**11 JAXA experiments performed since *Kibo* commissioning on 5/31/08

ESA	European Space Agency
Roscosmos	Russian Federal Space Agency
CSA	Canadian Space Agency
NASA	National Aeronautics and Space Administration
JAXA	Japanese Space Agency

In 2009, the number of astronauts living on board the ISS increased from three to six, and in 2010 the assembly of the ISS will be complete. As a result, more time will be spent on orbit performing ISS research. ISS laboratories are expected to accommodate an unprecedented amount of space-based research. Early utilization accomplishments give us hints about the value of a fully utilized ISS after assembly is complete.

(above) Astronaut works with the Smoke Point In Co-flow Experiment in the Microgravity Sciences Glovebox (MSG) during Expedition 18.

Cosmonaut performs inspection of the BIO-5 Rasteniya-2 (Plants-2) experiment in the Russian Lada greenhouse.

Plans Becoming a Reality

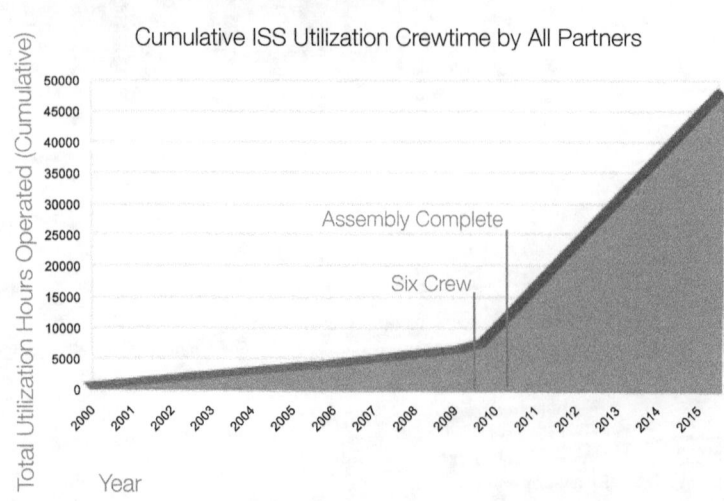

Cumulative ISS Utilization Crewtime by All Partners

Knowledge for all Humankind

Scientists from all over the world are already using ISS facilities, putting their talents to work in almost all areas of science and technology, and sharing their knowledge to make life on Earth better for people of all nations. We may not yet know what will be the most important knowledge gained from ISS, but we do know that there are some amazing discoveries on the way! Several recent patents and partnerships have already demonstrated Earth benefits of the public's investment in ISS research.

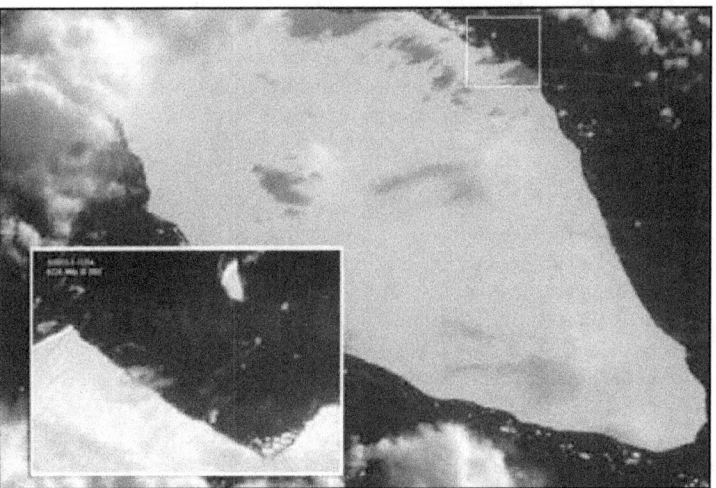

Regional view of Iceberg A22A, also known as "Amigosberg," with a detailed image of ice breakup along the margin. May 30, 2007.

Crew Earth Observations - International Polar Year (CEO-IPY) supported an international collaboration of scientists studying the Earth's Polar Regions from 2007 to 2009. Space station crew members photographed polar phenomena including icebergs, auroras, and mesospheric clouds. Observations, through digital still photography and video, from the ISS are used in conjunction with data gathered from satellites and ground observations to understand the current status of the Polar Regions. ISS, as a platform for these observations, will contribute data that has not been available in the past and will set the precedent for future international scientific collaborations for Earth observations. The International Polar Year, which started in 2007 and extended through February 2009, is a global campaign to study the Earth's polar regions and their role in global climate change.

Lab-on-a-Chip Application Development-Portable Test System (LOCAD-PTS) is a handheld device for rapid detection of biological and chemical substances on surfaces aboard the space station. Astronauts swab surfaces within the cabin, mix swabbed material in liquid form to the LOCAD-PTS, and obtain results within 15 minutes on a display screen, effectively providing an early

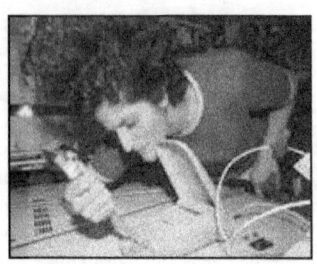

warning system to enable the crew to take remedial measures if necessary to protect themselves on board ISS. The handheld device is used with three different types of cartridges for the detection of endotoxin (a marker of gram-negative bacteria), glucan (fungi), and lipoteichoic acid (gram-positive bacteria). Lab-on-a-Chip technology has an ever-expanding range of applications in the biotech industry. Chips are available (or in development) that can also detect yeast, mold, and gram-positive bacteria, identify environmental contaminants, and perform quick health diagnostics in medical clinics.

Microbial Vaccine Development – Scientific findings from ISS research have shown increased virulence in Salmonella bacteria flown in space, and identified the controlling gene responsible. AstroGenetix, Inc. has funded their own follow-on studies on ISS and are now pursuing approval of a vaccine of an Investigational New Drug (IND) with the FDA. They are now applying a similar development approach to methycillin-resistant Staph aureus (MRSA).

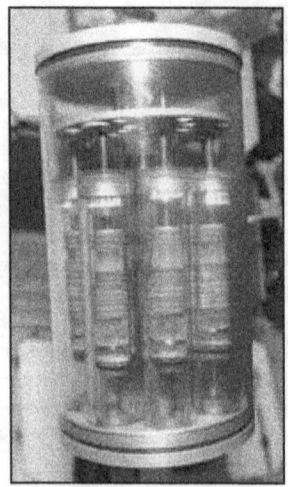

New Treatment Options for Duchenne Muscular Dystrophy: Collaborative High Quality Protein Crystal Growth – This JAXA and FSA-sponsored investigation was a unique collaboration between several ISS International Partners. The HQL-79 (human hematopoietic prostaglandin D2 synthase inhibitor) protein is a candidate treatment in inhibiting the effects of Duchenne muscular dystrophy. Investigators used the microgravity environment of the ISS to grow larger crystals and more accurately determine the three-dimensional structures of HQL-79 protein crystals. The findings led to the development of a more potent form of the protein, which is important for the development of a novel treatment for Duchenne muscular dystrophy. Russian investigators have collaborated internationally to grow macromolecular crystals on ISS since 2001, including genetically engineered human insulin (deposited into protein data bank in 2008), tuberculosis, and cholera-derived pyrophosphatase. The next generation of Russian-Japanese collaboration is the JAXA-High Quality Protein Crystal Growth experiment installed in *Kibo* in August 2009.

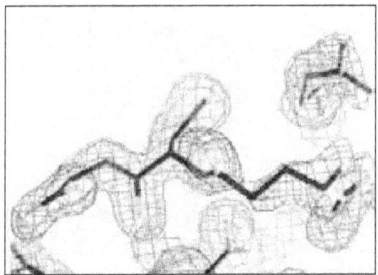

Electron density maps of HQL-79 crystals grown on Earth show a smaller three-dimensional structure (resolution of 1.7 Angstroms, top) as compared to the HQL-79 crystals grown in space (resolution of 1.28 Angstroms, bottom).

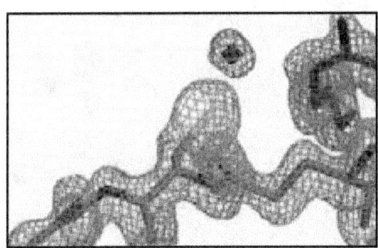

The **Plasma Crystal** experiment was one of the first scientific experiments performed on ISS in 2001. Complex plasma is a low-temperature gaseous mixture composed of ionized gas, neutral gas, and micron-sized particles. Under specific conditions, the interactions of these microparticles lead to a self-organized structure of a "plasma crystal" state of matter. Gravity causes the microparticles to sediment due to their relatively high mass compared to that of the ions, and so they have to be electrostatically levitated for proposer development. The microgravity environment of the ISS allowed the development of larger three-dimensional plasma crystal systems in much weaker electric fields than those necessary for the levitation on the ground, revealing unique structural details of the crystals. The ESA is now building the next generation of complex plasma experiments for the ISS in collaboration with a large international science team. Understanding the formation and structure of these plasma crystal systems can also lead to improvements in industrial process development on Earth.

Plasma Crystal 3 Plus [Roscosmos, DLR, ESA], as well as previous experiments of this series, is one example of a complex set of plasma crystal experiments that allow scientists to study crystallization and melting of dusty plasma in microgravity by direct viewing of those phenomenon. The equipment includes a tensor unit, turbo pump, and two TEAC Aerospace Technologies video tape recorders are part of the telescience equipment. Video recordings of the plasma crystal formation process, along with parameters such as gas pressure, high-frequency radiated power and the size of dust particles are downlinked to Earth for analysis.

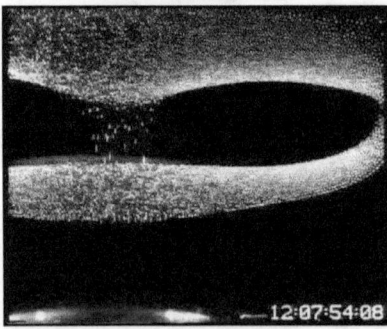

Dusty Plasma in Microgravity

This challenging complex plasma science program has been ongoing for many years and is another example of fruitful collaboration between multiple European countries and Russia, Japan and the U.S. It is a very long-term effort and will require at least another 5-10 years of intensive research on ISS.

Knowledge for all Humankind

An ISS investigator recently patented the **Microparticle Analysis System and Method**, an invention for a device that detects and analyzes microparticles. This technology supports the chemical and pharmaceutical industries, and is one of a sequence of inventions related to technology development for experiments on the ISS and shuttle, including the Microencapsulation Electrostatic Processing System (MEPS) experiment that demonstrated microencapsulation processing of drugs, a new and powerful method for delivering drugs to targeted locations. MEPS technologies and methods have since been developed that will be used to deliver microcapsules of anti-tumor drugs directly to tumor sites as a form of cancer therapy.

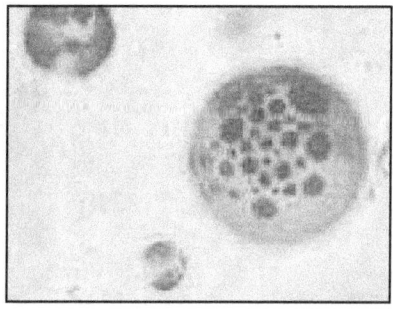

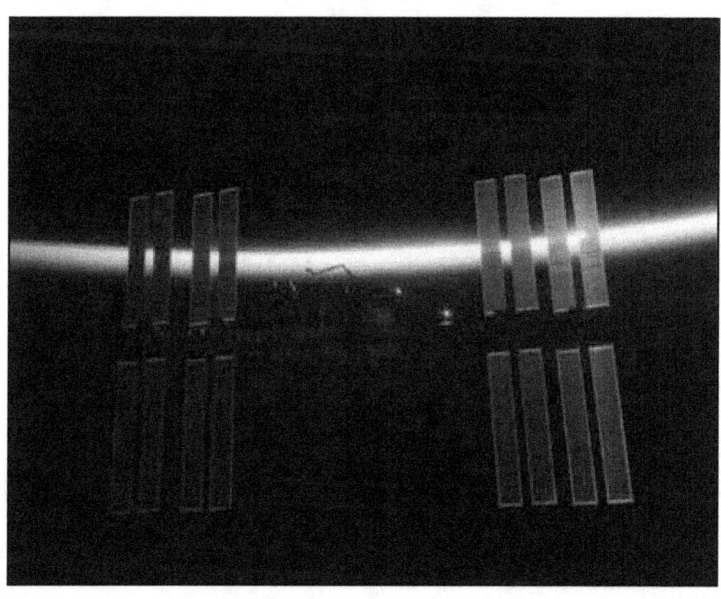

Advanced Diagnostic Ultrasound in Microgravity (ADUM) – The ultrasound is the only medical imaging device currently available on ISS. This experiment demonstrated the diagnostic accuracy of ultrasound in medical contingencies in space and determined the ability of minimally trained crew members to perform ultrasound examinations with remote guidance from the ground. The telemedicine strategies investigated by this experiment could have widespread application and have been applied on Earth in emergency and rural care situations. In fact, the benefits of this research are being used in professional and amateur sports from hockey, baseball, and football teams, to the U.S. Olympic committee. Sport physicians and trainers can now perform similar scans on injured players at each of their respective sport complexes by taking advantage of ultrasound experts available remotely at the Henry Ford Medical System in Detroit. This is an excellent example of how research aboard the ISS continues to be put to good use here on Earth while, at the same time, paving the way for our future explorers.

An Unprecedented International Laboratory

The laboratories of the ISS are virtually complete; key research facilities – science laboratories in space – are up and running. In 2008, the ESA *Columbus* and JAXA *Kibo* laboratories joined the U.S. *Destiny* Laboratory and the Russian *Zvezda* Service Module. *Zvezda* was intended primarily to support crew habitation, but became the first multipurpose research laboratory of ISS. In addition, the U.S. has expanded its user base beyond NASA to other government agencies and the private sectors to make ISS a U.S. National Laboratory. Additional science facilities will travel to the ISS on the remaining assembly missions in 2009 and 2010.

As all ISS partner nations begin their research programs, international collaboration and interchange among scientists worldwide is growing rapidly. Over the final years of assembly in 2009-2010, the initial experiments have been completed in the newest racks, the crew size on board ISS has doubled to six astronauts/cosmonauts, and in 2010 we will transition from "early utilization" to "full utilization" of ISS. The ISS labs are GO!

This high-flying international laboratory is packed with some of the most technologically sophisticated facilities that can support a wide range of scientific inquiry in biology, human physiology, physical and materials sciences, and Earth and space science. There is probably no single place on Earth where you can find such a laboratory – approximately the size of a U.S football field (including the end zones) and has the interior volume of 1.5 Boeing 747 jetliners – with facilities to conduct the breadth of research that can be done aboard the ISS. Keep turning the pages to learn more about this amazing laboratory orbiting approximately 350 km (220 miles) above us.

ISS Laboratory Research Rack Locations at Assembly Complete

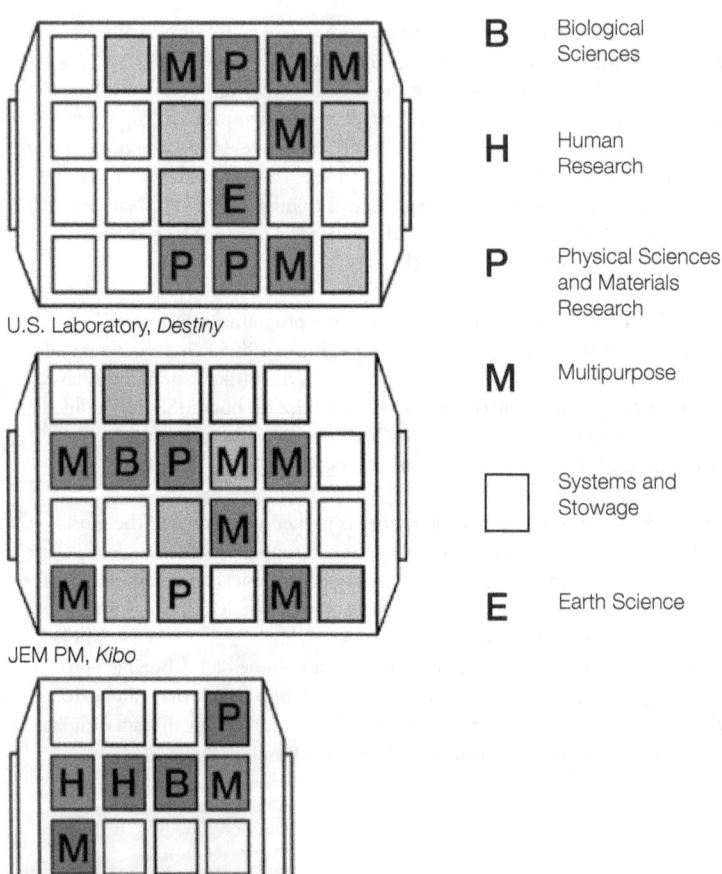

ISS Payload Accommodations

International Standard Payload Rack (ISPR) Sites

Power
3, 6, or 12 kw, 114.5-126 VDC

Data
Lo Rate: MIL-STD- 1553 bus
1 Mbps
High Rate: 100 Mbps
Ethernet: 10 Mbps
Video: NTSC

Gases
Nitrogen
Flow=0.1 kg/minute minimum
517-827 kPa nominal
1379 kPa max
Argon, Carbon Dioxide, Helium
517-768 kPa nominal
1379 kPa maximum

Cooling Loops
Moderate Temperature:
16.1°C - 18.3°C
Flow rate=0-45.36 kg/hr
Low Temperature:
3.3°C - 5.6°C
Flow rate=233 kg/hr

Vacuum
Venting: 10-3 torr in less than 2 hours
Vacuum Resource: 10^{-3} torr

Distributed payload service in Zvezda and Pirs provide direct connection to electrical, thermal, and vacuum.

ISS External Payload Sites

EXPRESS Logistics Carrier (Truss)
Mass: 4445.2 kg
Volume: 30 m3
Power: 3 kW max, 113-126 VDC
Data: Low Rate: 1 Mbps MIL-STD-1553
High Rate: 95 Mbps (shared)

EXPRESS Adapter Site
Mass: 226.8 kg
Volume: 1 m3
Power: 750 W max, 113-126 VDC, 500 W max, 28 VDC
Data: Low Rate: 1 Mbps MIL-STD-1553
Medium Rate: 6 Mbps (shared)

Japanese Experiment Module (JEM) Exposed Facility (EF) Site
Mass: 521.63 kg Standard Site
2494.8 kg Large Site
Volume: 1.5 m3
Power: 3 kW max, 113-126 VDC
Data: Low Rate: 1 Mbps MIL-STD-1553
High Rate: 43 Mbps (shared)
Ethernet: 10 Mbps

ColumbusExternal Payload Facility (CEPF) Site
Mass: 226.8 kg
Volume: 1 m3
Power: 2.5 kW max, 120 VDC (shared)
Data: Low Rate: 1 Mbps MIL-STD-1553
Medium Rate: 2 Mbps
Ethernet: 10 Mbps

Laboratory Facilities

18

Multipurpose Facilities

European Drawer Rack (EDR) [ESA] is a multidiscipline facility to support up to seven modular experiment modules. Each payload will have its own cooling, power, and data communications, vacuum, venting, and nitrogen supply. EDR facilitates autonomous operations of subrack type of experiments in a wide variety of scientific disciplines.

The Protein Crystallization Diagnostics Facility (PCDF) is the first ESA experiment performed with the EDR rack. Its main science objectives are to study the protein crystal growth conditions by way of non-intrusive optical techniques like Dynamic Light Scattering (DLS), Mach-Zehnder Interferometry (MZI) and classical microscopy. Understanding how crystals grow in purely diffusive conditions helps define the best settings to get organic crystals as perfect as possible. Later on these crystals will be preserved and analyzed via X-rays on Earth to deduce the three-dimensional shape of proteins.

Multipurpose Small Payload Rack (MSPR) [JAXA] has two workspaces and one workbench and can hold equipment, supply power, and enable communication and video transmission. With such general characteristics, MSPR can be used in various fields of space environment use not only for science, but also for cultural missions.

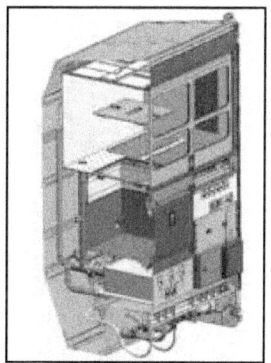

Expedite the Processing of Experiments to Space Station – (EXPRESS) Racks [NASA] are modular multipurpose payload racks that store and support experiments aboard ISS. The rack provides structural interfaces, power, data, cooling, water, and other items needed to operate the science experiments on ISS. Experiments are exchanged in and out of the EXPRESS Rack as needed; some subrack multi-user facilities (like European Modular Cultivation [EMCS]) will remain in EXPRESS for the life of ISS, while others are used for only a short period of time.

Multipurpose Facilities

Multipurpose Facilities

EXPRESS Rack Designs

Over 50% of the capabilities of EXPRESS Racks are available for new research equipment. EXPRESS Racks are the most flexible modular research facility available on ISS and are used for NASA and international cooperative research.

EXPRESS 2 - *Destiny*

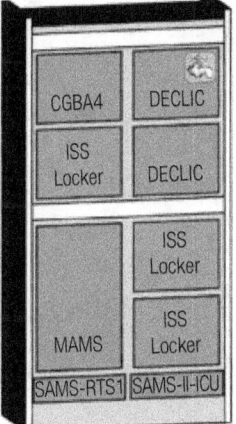

EXPRESS 2A - *Destiny*

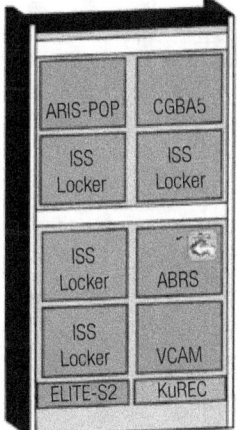

EXPRESS 3A - *Columbus*

EXPRESS 5 - *Kibo*

Multipurpose Facilities

EXPRESS 6 - *Destiny*

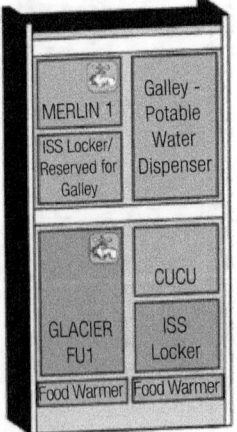

EXPRESS 7 - *Destiny*

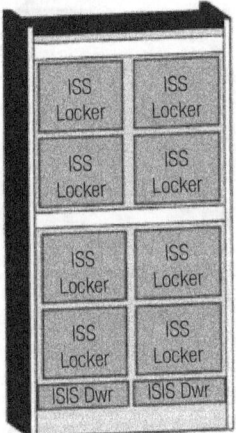

EXPRESS 4 - *Kibo*

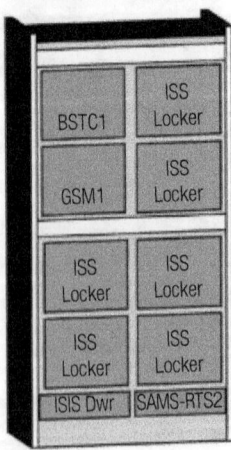

 Available for future utilization

 Facilities currently in use

 Systems Hardware

 Water-Cooled Payloads

Multipurpose Facilities

Freezers *allow freezing, storage, and transportation of science samples collected on ISS for later return to Earth. Although the original plan was to launch only one **MELFI** Freezer for ISS, NASA is now launching three **MELFI** freezers to support physiology and life science research by the U.S. National Laboratory users and all of the ISS partners.*

General Laboratory Active Cryogenic ISS Equipment Refrigerator (GLACIER) [NASA] serves as an on-orbit ultra-cold freezer (as low as -165°C) and has a volume of 11.35 L.

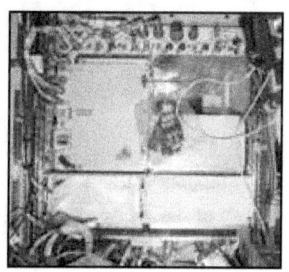

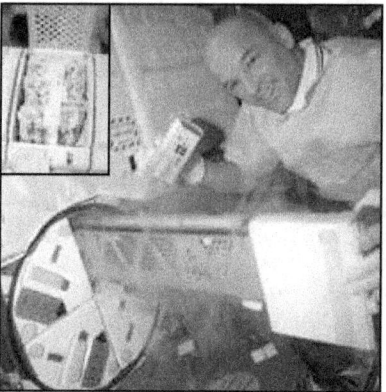

Minus Eighty-Degree Laboratory Freezer for ISS (MELFI) [ESA, NASA] is a refrigerator/freezer for biological and life science samples collected on ISS. These ESA-built and NASA-operated freezers store samples at temperatures of +4°C to as low as –80°C and each has a volume of 175 L of samples.

Microgravity Experiment Research Locker/Incubator (MERLIN) [NASA] can be used as either a freezer, refrigerator, or incubator (between -20.0°C to 48.5°C) and has a volume of 4.17 L.

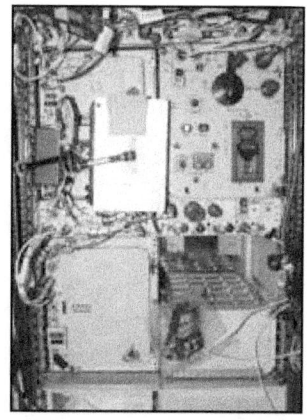

Microgravity Science Glovebox (MSG) [ESA, NASA] provides a safe environment for research with liquids, combustion, and hazardous materials on board ISS. Crew members access the work area through ports equipped with rugged, sealed gloves. A video system and data downlinks allow for control of the enclosed experiments from the ground. Built by ESA and operated by NASA, MSG is the largest glovebox flown in space.

Multipurpose Facilities

Gloveboxes provide containment of experiments, ensuring that hazardous materials do not float about the cabin. The MSG has been the most heavily used facility during ISS construction. In one short period in 2008, it was used for a combustion experiment, a study of complex fluids, and to harvest plants. Again in 2009 a wide variety of experiments will be using the versatile MSG accommodation and functional capabilities.

Portable Glove Box (PGB) [ESA] is a small glovebox that can be transported around ISS and used to provide two levels of containment for experiments in any laboratory module. Three levels of containment can be achieved by placing the PGB inside the larger volume of the MSG (above).

Biological Research

Advanced Biological Research System (ABRS) [NASA] is a single locker system with two growth chambers. Each growth chamber is a closed system capable of independently controlling temperature, illumination, and atmospheric composition to grow a variety of biological organisms including plants, microorganisms, and small arthropods (insects and spiders).

*The first plant experiments in **ABRS** will include the first trees flown in space (willows for a Canadian study of cambium formation) and an American study will use green fluorescent proteins as environmental stress indicators.*

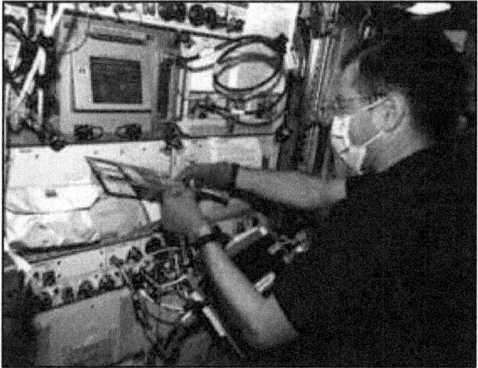

Biotechnology Specimen Temperature Controller (BSTC) [NASA] includes a refrigerator, incubator, and cryo-drawer, as well as envrionmental and atmospheric control to grow and maintain mammalian cell cultures in microgravity.

Biological Laboratory (BioLab) [ESA] is used to perform space biology experiments on microorganisms, cells, tissue cultures, small plants, and small invertebrates, and will allow a better understanding of the effects of microgravity and space radiation on biological organisms. BioLab includes an incubator, with a microscope, spectrophotometer, and two centrifuges to provide artificial gravity. It also has a glovebox and two cooler/freezer units.

Waving and Coiling of Arabidopsis Roots at Different g-levels (WAICO) was the first experiment conducted in BioLab. Plant growth is impacted by several factors; i.e., temperature, humidity, gravitropism, phototropism, and circumnutation. Shoots/stems and roots develop following complex phenomena at micro-/macroscopic levels. The goal of this experiment was to understand the interaction of circumnutation (the successive bowing or bending in different directions of the growing tip of the stems and roots) and gravitropism (a tendency to grow toward or away from gravity) in microgravity and 1-g of *Arabidopsis thaliana* wild type and an agravitropic mutant. A follow-on study is planned for 2010.

Exposure Experiment (Expose) [ESA] is a multi-user facility accommodating experiments in the following disciplines: photo processing, photo-biology and exobiology. Expose allows short- and long-term exposure of experiments to space conditions and solar UV radiation on the ISS. The Expose facilities are installed on the external surfaces of *Zvezda* service module and *Columbus* module.

Biological Research Facilities

Biological Research Facilities

Commercial Generic Bioprocessing Apparatus (CGBA) [NASA] provides programmable, accurate temperature control – from cold stowage to a customizable incubator – for experiments that examine the biophysical and biochemical actions of microorganisms in microgravity. CGBA can be used in a wide variety of biological studies, such as protein crystal growth, small insect habitats, plant development, antibiotic-producing bacteria, and cell culture studies.

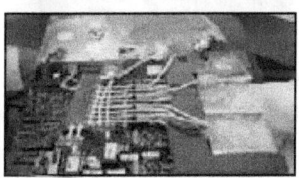

*CGBA, operated by **Bioserve Space Techologies**, is a key facility being used by U.S. investigators as part of the ISS National Laboratory initiative.*

European Modular Cultivation System (EMCS) [ESA, NASA] allows for cultivation, stimulation, and crew-assisted operation of biological experiments under well-controlled conditions (e.g., temperature, atmospheric composition, water supply, and illumination). It is being used for multi-generation experiments and studies of gravitational effects on early development and growth in plants and other small organisms.

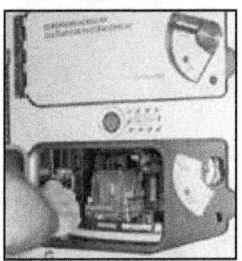

*The **EMCS** has two centrifuges that can spin at 0 to 2x Earth's gravity. Different experiment containers can hold a variety of organisms, such as worms and fruit flies, as well as seeds and plants. The EMCS has already supported a number of plant growth experiments operated by ESA, NASA, and JAXA.*

Kriogem-3M [Roscosmos] is a refrigerator-incubator used for stowage of biological samples, and for the culture and incubation of bioreactors such as **Recomb-K**. Bioreactors are specialized hardware for growing cells, tissues, and microorganisms.

LADA Greenhouse **[Roscosmos]** – Since its launch in 2002, the LADA greenhouse has been in almost continous use for growing plants in the Russian segment. It has supported a series of experiments on fundamental plant biology and space farming, growing multiple generations of sweet peas, wheat, tomatoes, and lettuce.

*NASA and Roscosmos have used the **LADA** greenhouse in cooperative tests to determine the best ways to keep roots moist in space. Bioregenerative life support from photosynthesis may be an important component of future spacecraft systems.*

Biological Research Facilities

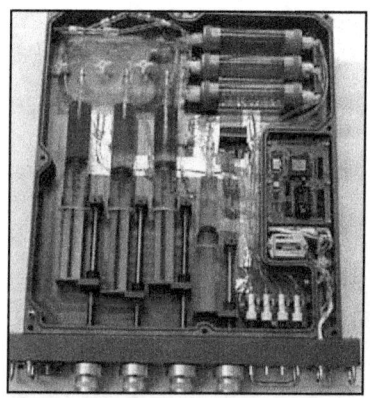

Eosteo Bone Culture System [CSA] provides the right conditions to grow bone cells in microgravity. This culture system has been used successfully on U.S. Space Shuttle and Russian Foton recoverable orbital flights, and is also available for use in bone cell culture on ISS.

Understanding the cellular changes in bone cells in orbit could be key for understanding the bone loss that occurs in astronauts while they are in space.

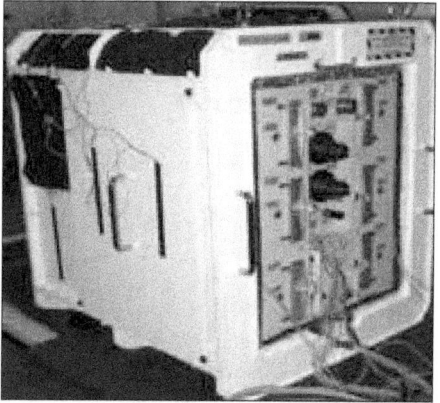

Mice Drawer System (MDS) [NASA, ASI] is hardware provided by the Italian Space Agency that will use a validated mouse model to investigate the genetic mechanisms underlying bone mass loss in microgravity. MDS is a multifunctional and multiuser system that allows experiments in various areas of biomedicine, from research on organ function to the study of the embryonic development of small mammals under microgravity conditions. Research conducted with the MDS is an analogue to the human research program, which has the objective to extend the human presence safely beyond low-Earth orbit.

Saibo Experiment Rack (Saibo) [JAXA] is a multipurpose payload rack system that sustains life science experiment units inside and supplies resources to them. Saibo consists of a Clean Bench, a glovebox with a microscope, and a **Cell Biology Experiment Facility (CBEF),** which has incubators, a centrifuge, and sensors to monitor the atmospheric gases.

Saibo means "living cell." The first use of Saibo was for studies of the effects of radiation on immature immune cells.

Aquatic Habitat (AQH) [JAXA] enables breeding experiments with medaka or zebrafish in space, and those small freshwater fish have many advantages as one of the model animals for study. The AQH is composed of two aquariums, which have automatic feeding systems, LED lights to generate day/night cycle, and Charge-coupled device (CCD) cameras for observation.

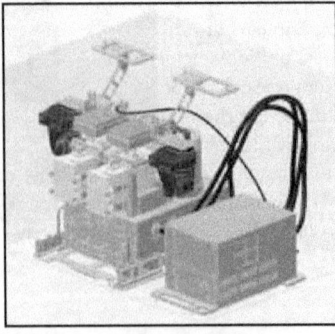

Human Physiology Research

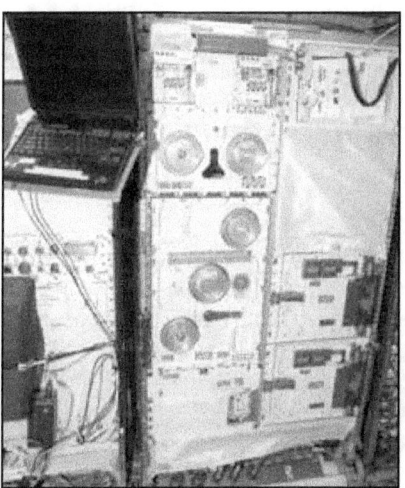

European Physiology Module (EPM) [ESA] is designed for investigating the effects of microgravity on short-term and long-duration spaceflights on the human body and includes equipment for studies in neuroscience, cardiovascular, bone and muscle physiology, as well as investigations of metabolic processes. The cardiolab instrument was provided by CNES and DLR.

Muscle Atrophy Research Exercise System (MARES) [ESA] will be used for research on musculoskeletal, biomechanical, and neuromuscular human physiology to better understand the effects of microgravity on the muscles. This instrument is capable of assessing the strength of isolated muscle groups around joints by controlling and measuring relationships between position/velocity and torque/force as a function of time.

Human Research Facility (HRF-1 and HRF-2) [NASA] enables human life science researchers to study and evaluate the physiological, behavioral, and chemical changes induced by long-duration spaceflight. HRF-1 houses medical equipment including a **Clinical Ultrasound**, **Space Linear Acceleration Mass Measurement Device (SLAMMD)** for measuring on-orbit crew member mass, devices for measuring **Blood Pressure** and **Heart Function**, and a **Refrigerated Centrifuge** for processing blood samples. The equipment is being used to study the effects of long-duration spaceflight on the human body. Researchers will use ISS to understand the physiology and to test countermeasures that will prevent negative effects of space travel, and enable humans to travel beyond Earth orbit.

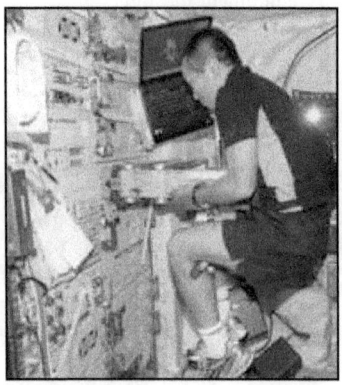

SLAMMD

Techniques developed for using ultrasound technology on ISS are now being used in trauma facilities to more rapidly assess serious patient injuries (also see page 14).

Refrigerated Centrifuge

Clinical Ultrasound

Human Physiology Research Facilities

Human Physiology Research Facilities

Pulmonary Function System (PFS) [ESA, NASA] is hardware developed collaboratively by ESA and NASA. It includes four components that are needed to make sophisticated studies of lung function by measuring respired gases in astronaut subjects. It includes two complimentary analyzers to measure the gas composition of breath, the capability to make numerous different measurements of lung capacity and breath volume, and a system to deliver special gas mixtures that allow astronauts to perform special tests of lung performance. ESA will also be operating a small portable version of the system (Portable PFS) that can be used in the various laboratory modules.

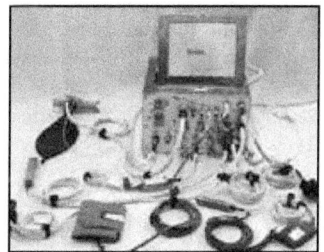

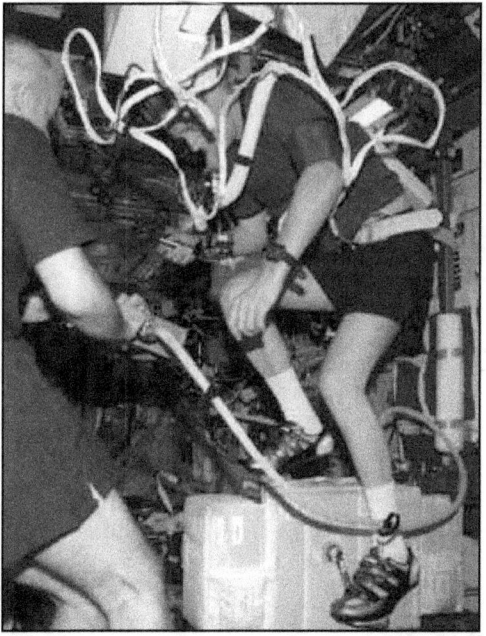

SLAMMD and *PFS* are used by flight surgeons during periodic medical exams on ISS. Understanding the gradual deconditioning of astronauts and cosmonauts during their stay on ISS is critical for developing better exercise capabilities for exploration beyond Earth orbit.

Human Research Hardware [CSA] is used cooperatively with other international hardware for better understanding of the physiological responses to spaceflight. The hardware includes radiation dosimeters (EVARM), and hardware and software for studying hand-eye coordination and visual perception (Perceptual Motor Deficits in Space [PMDIS], Bodies In the Space Environment [BISE]) and neurophysiology (H-REFLEX).

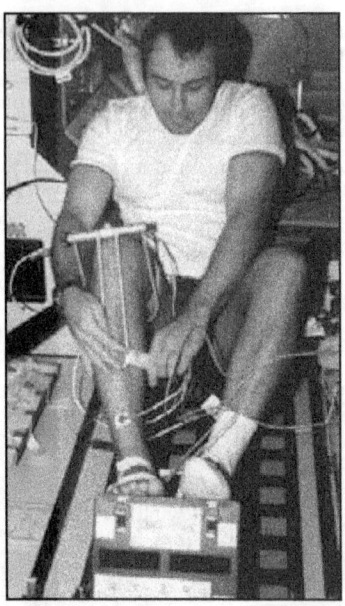

H-REFLEX

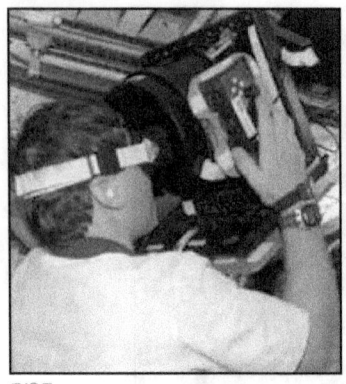

BISE

PMDIS

EVARM

Human Physiology Research Facilities

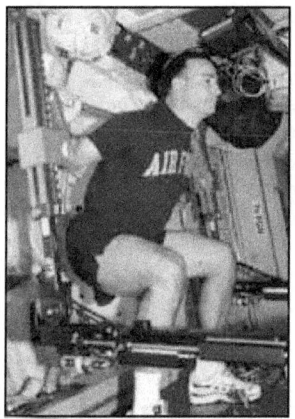

The **Advanced Resistive Exercise Device (ARED) [NASA]** is systems hardware that provides exercise capabilities to crew members on the ISS. The ARED also collects data regarding the parameters (loads, repetitions, stroke, etc.) associated with crew exercise, and transmits it to the ground.

The **Cycle Ergometer with Vibration Isolation System (CEVIS) [NASA]** provides the ability for recumbent cycling to provide aerobic exercise as a countermeasure to cardiovascular deconditioning on orbit.

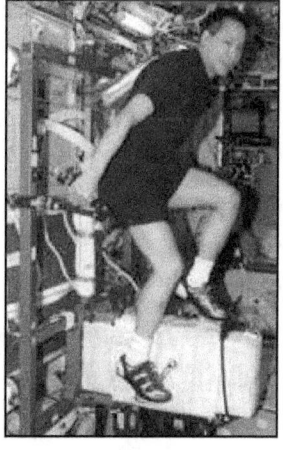

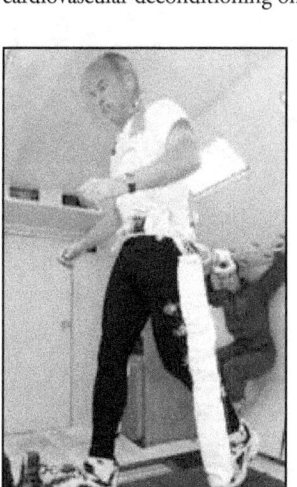

The **Combined Operational Load Bearing External Resistive Exercise Treadmill (COLBERT) [NASA]** can collect data such as body loading, duration of session, and speed for each crew member.

The second generation of exercise equipment used for daily exercise on board ISS collects information on protocols and forces that are used as supplemental data for studies of muscle and bone loss and cardiovascular health during long-duration spaceflight.

Measuring Radiation Hazards in Space (Matryoshka) [ESA, Roscosmos, NASA, JAXA] is a series of investigations to measure radiation doses experienced by astronauts in space outside (MTR-1) and at various locations inside (MTR-2) the ISS. Matryoshka uses a mannequin of a human torso made of plastic, foam, and a real human skeleton. The torso is equipped with dozens of radiation sensors that are placed in strategic locations throughout its surface and interior to measure how susceptible different organs and tissue may be to radiation damage experienced by astronauts in space. Research institutes from around the world have collaborated and shared data from the project. The results will give the radiation dose distribution inside a human phantom torso for a better correlation between skin and organ dose and for better risk assessment in future long-duration spaceflight.

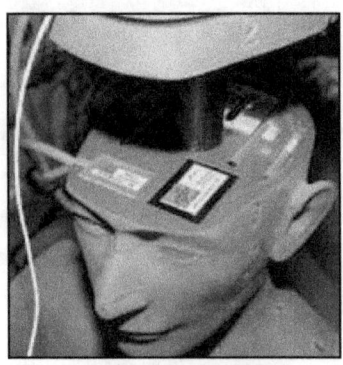

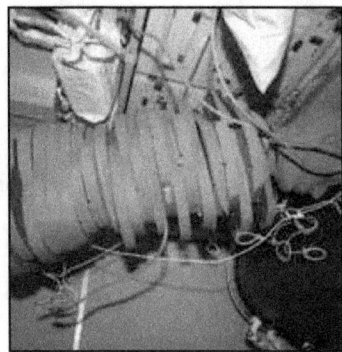

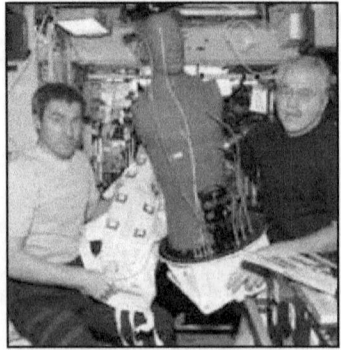

Participants from 10 countries provided dosimeters and other components of Matryoshka, making it one of the largest multinational collaborative investigations on the ISS. The Matryoshka program started in 2004 and will incrementally continue for some years.

Human Physiology Research Facilities

Human Life Research [Roscosmos] includes a variety of devices and systems designed to study human life in space. Components of the system of equipment include the Cardiovascular System Research Rack, Weightlessness Adaptation Study Kit, Immune System Study Kit, and Locomotor System Study Facility.

Weightless Adaptation

Locomotor System

Human Research Hardware [JAXA] includes a portable **Digital Holter ECG** recorder for 24-hour electrocardiogram monitoring of cardiovascular and autonomic function of the astronauts. The recorded data are downlinked through the Multi-protocol Converter, and **Crew PADLES**, which is a passive dosimeter that records the personal dose of the astronauts. The dose records are used to assess a radiation exposure limit of each astronaut.

Human physiology research is coordinated by an internal working group to coordinate experiments and share data. An astronaut or cosmonaut can participate in as many as 20 physiology experiments during their stay on ISS.

PADLES

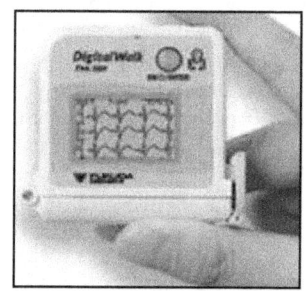
Holter

Human Physiology Research Facilities

Anomalous Long Term Effects in Astronaut's Central Nervous System (ALTEA) [ASI, NASA, ESA] ALTEA is a helmet-shaped device holding six silicon particle detectors that has been used to measure the effect of the exposure of crew members to cosmic radiation on brain activity and visual perception, including astronauts' perceptions of light flashes behind their eyelids as a result of high energy radiation. Because of its ability to be operated without a crew member, it is also being used as a portable dosimeter to provide quantitative data on high energy radiation particles passing into ISS.

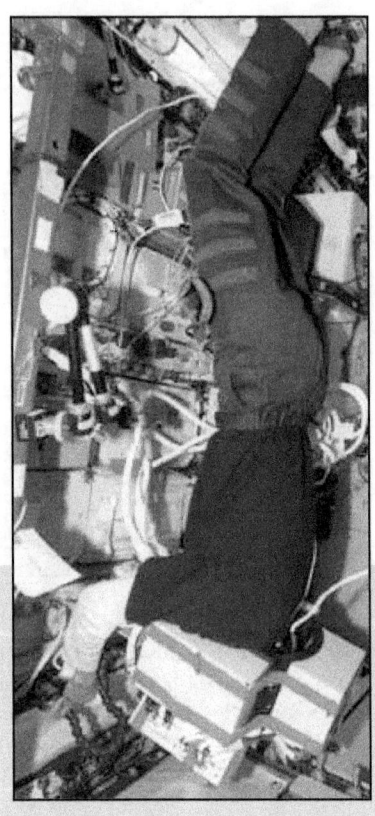

ALTEA-Dosi capabilities are also used to give additional information on exposure of crew members to radiation during their stays on ISS for use in health monitoring. ALTEA-Shield will provide data about radiation shielding effects by a variety of special materials.

Hand Posture Analyser (HPA) [NASA, ASI] is composed of the Handgrip Dynamometer / Pinch Force Dynamometer, the Posture Acquisition Glove and the Inertial Tracking System (ITS) for the measurement of finger position and upper limb kinematics. The HPA examines the way hand and arm muscles are used differently during grasping and reaching tasks in weightlessness.

Human Physiology Research Facilities

ELaboratore Immagini Televisive-Space 2 (ELITE-S2) [NASA, ASI] investigates the connection between brain, visualization, and motion in the absence of gravity. By recording and analyzing the three-dimensional motion of astronauts, this study will help engineers apply ergonomics into future spacecraft designs and determine the effects of weightlessness on breathing mechanisms for long-duration missions. Results might also be applied to neurological patients on the ground with impaired motor control.

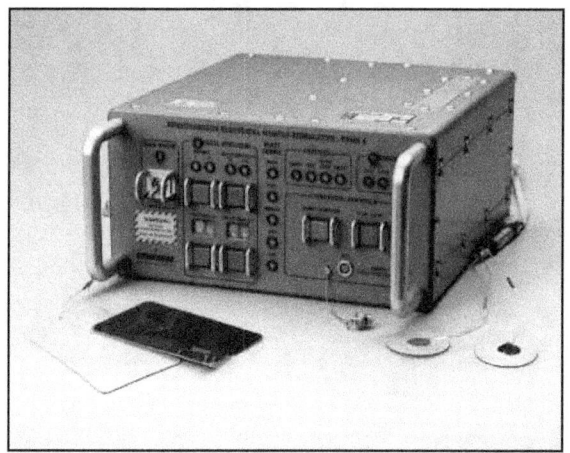

Percutaneous Electrical Muscle Stimulator (PEMS) [ESA] is a self-contained, locker stowed item. Its purpose is to deliver electrical pulse stimulation to non-thoracic muscle groups of the human test subject, thereby creating contractile responses from the muscles. The PEMS supports neuromuscular research. It provides single pulses or pulse trains according to a preadjusted program.

Frank De Winne
Belgium
ESA

Roman Romanenko
Russia
Roscosmos

Gennady Padalka
Russia
Roscosmos

Michael Barratt
United States
NASA

Robert Thirsk
Canada
CSA

Koichi Wakata
Japan
JAXA

Human Physiology Research Facilities

Expedition 20 represented a milestone on board the ISS. It was the first time each international partner had a representative on board the station at the same time.

Physical Science and Materials Research

Combustion Integrated Rack (CIR) [NASA] is used to perform sustained, systematic combustion experiments in microgravity. It consists of an optics bench, a combustion chamber, a fuel and oxidizer management system, environmental management systems, interfaces for science diagnostics and experiment specific equipment, as well as five different cameras to observe the patterns of combustion in microgravity for a wide variety of gasses and materials.

The *Multi-User Droplet Combustion Apparatus - Flame Extinguishment Experiment* creates droplets of fuel that ignite while suspended in a containment chamber.

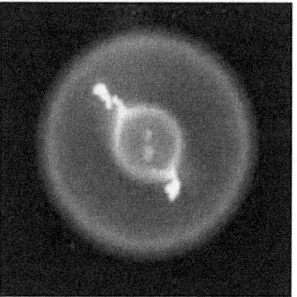

Example of a burning droplet from a previous space combustion experiment.

Fluid Science Laboratory (FSL) [ESA] is a multi-user facility for conducting fluid physics research in microgravity conditions. The FSL provides a central location to perform fluid physics experiments on board ISS that will give insight into the physics of fluids in space, including aqueous foams, emulsions, convection, and fluid motions. Understanding how fluids behave in microgravity will lead to development of new fluid delivery systems in future spacecraft design and development.

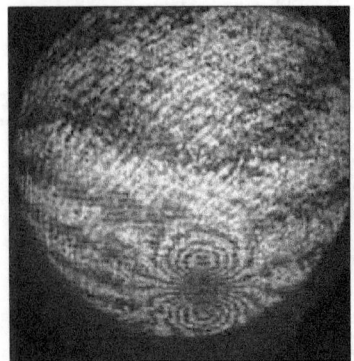

Geoflow Interferogram Image.

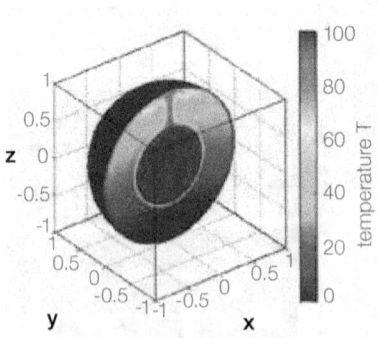

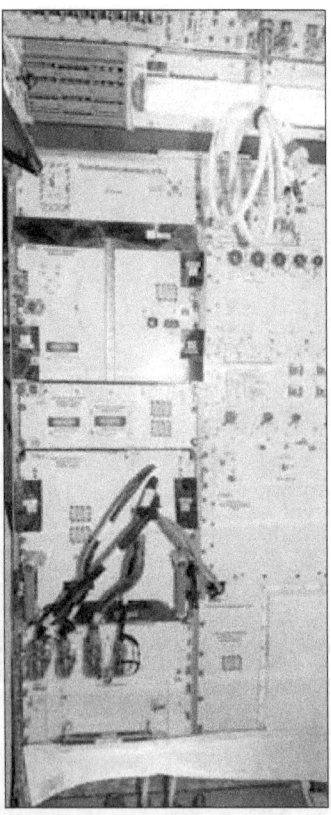

GEOFLOW was the first Experiment Container processing FSL. The first experiment in the FSL studied a model of liquid core planets.

Physical Science and Materials Research Facilities

Fluids Integrated Rack (FIR)

[NASA] is a complementary fluid physics research facility designed to accommodate a wide variety of microgravity fluid experiments and the ability to image these experiments. FIR features a large user-configurable volume for experiments. The FIR provides data acquisition and control, sensor interfaces, laser and white light sources, advanced imaging capabilities, power, cooling, and other resources. The FIR will host fluid physics investigations into areas such as complex fluids (colloids, gels), instabilities (bubbles), interfacial phenomena (wetting and capillary action), and phase changes (boiling and cooling). Fluids under microgravity conditions perform differently than those on Earth. Understanding how fluids react in these conditions will lead to improved designs on fuel tanks, water systems, and other fluid-based systems.

*The **FIR** includes the Light Microscopy Module (LMM). The LMM is a remotely controllable (commanded from the ground), automated microscope that allows flexible imaging (bright field, dark field, phase contrast, etc.) for physical and biological experiments.*

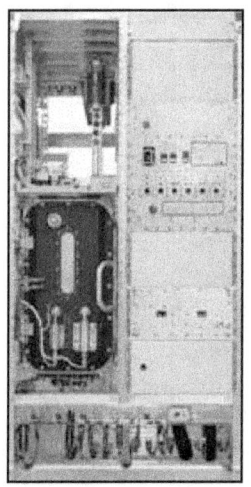

Kobairo Rack with Gradient Heating Furnace (GHF) [JAXA] is an electrical furnace to be used for generating high-quality crystals from melting materials. It consists of a vacuum chamber and three independently movable heaters, which can realize high temperature gradient up to 150°C/cm.

Materials Science Research Rack (MSRR-1) [ESA, NASA] will provide a powerful, multi-user **Materials Science Laboratory (MSL)** in the microgravity environment of ISS and can accommodate studies of many different types of materials. Experiment modules that contain metals, alloys, polymers, semiconductors, ceramics, crystals, and glasses can be studied to discover new applications for existing materials and new or improved materials (crystal growth, longer polymer chains, and purer alloys). MSRR will enable this research by providing hardware to control the thermal, environmental, and vacuum conditions of experiments, monitoring experiments with video, and supplying power and data handling for specific experiment instrumentation.

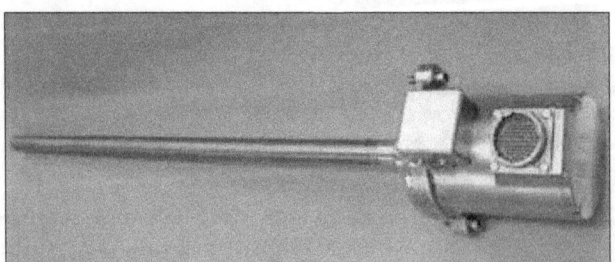

Sample Cartridge Assembly

*Experiments in the **MSRR** are coordinated by international teams that share different parts of the samples. There are 25 investigators on three research teams participating in the first of these investigations. MSL – Columnar-to-Equiaxed Transition in Solidification Processing and Microstructure Formation in Casting of Technical Alloys under Diffusive and Magnetically Controlled Convective Conditions (MSL-CETSOL and MICAST) are two investigations that support research into metallurgical solidification, semiconductor crystal growth (Bridgman and zone melting), and measurement of thermo-physical properties of materials.*

Physical Science and Materials Research Facilities

Physical Science and Materials Research Facilities

Ryutai Experiment Rack (Ryutai) [JAXA] is a multipurpose payload rack system that supports various fluid physics experiments. Ryutai consists of four sub-rack facilities: **Fluid Physics Experiment Facility (FPEF)**; **Solution Crystallization Observation Facility (SCOF)**; **Protein Crystallization Research Facility (PCRF)**; and **Image Processing Unit (IPU)**. Ryutai enables teleoperations of the experiments providing the electrical power, ground command and telemetry monitoring, water cooling, and gas supply to those sub-rack facilities.

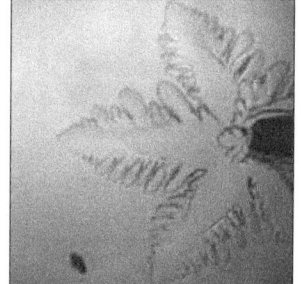

Ryutai means "fluid." The JAXA experiment Ice Crystal examines the factors that lead to the pattern formation in ice crystals in microgravity.

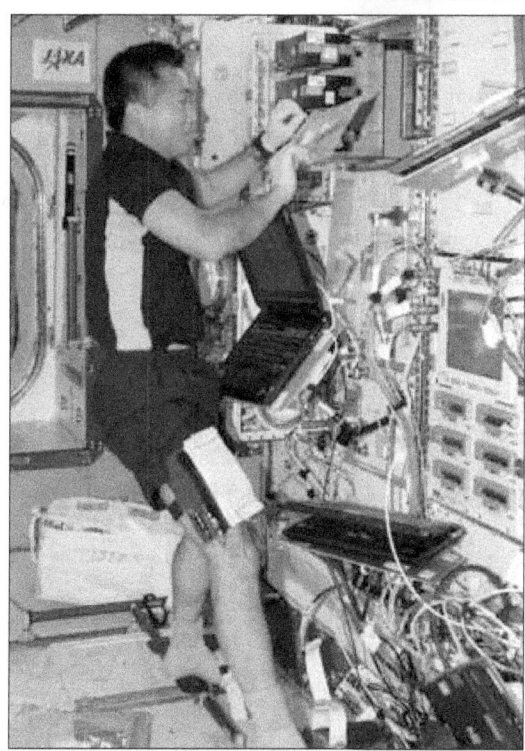

Space Dynamically Responding Ultrasonic Matrix System (SpaceDRUMS) [NASA] will provide a suite of hardware capable of facilitating containerless advanced materials science, including combustion synthesis and fluid

physics. SpaceDRUMS uses ultrasound to completely suspend a baseball-sized solid or liquid sample during combustion without the materials ever contacting the container walls. Such advanced ceramics production may have applications in new spacecraft or extraterrestrial outposts, such as bases on the moon.

Device for the study of Critical Liquids and Crystallization (DECLIC) [CNES, NASA] is a multi-user facility developed by the ESA-member agency Centre National d'Etudes Spatiales (French Space Agency, [CNES]) and flown in collaboration with NASA. It was designed to conduct experiments in the fields of fluid physics and materials science. A special insert allows the study of both ambient temperature critical point fluids and high-temperature super-critical fluids. Another class of insert will study the dynamics and morphology of the fronts that form as a liquid material solidifies.

Materials International Space Station Experiment (MISSE) [NASA] is a series of external exchangeable test beds located on ESA's Columbus-EPF for studying the durability of materials such as optics, sensors, electronics, communications devices, coatings, and structural materials. To date, a total of six different MISSE experiments have been attached to the outside of the ISS and evaluated for the effects of atomic oxygen, vacuum, solar radiation, micrometeorites, direct sunlight, and extremes of heat and cold. This experiment allows the development and testing of new materials to better withstand the rigors of space environments. Results will provide a better understanding of the durability of various materials when they are exposed to such an extreme environment. Many of the materials may have applications in the design of future spacecraft.

Results from **MISSE** tests have led to changes in materials used in dozens of spacecraft built over the last 5 years.

Replaceable Cassette-Container (SKK or CKK) [Roscosmos] is mounted on the outside of ISS to test materials that are directly exposed to the harsh environment of space. CKK are detachable cassette containers that measure the level and composition of contamination and monitor the change in operating characteristics for samples of materials from the outside surfaces of the ISS Russian segment. The CKK is a two-flap structure and consists of a casing and spool holders containing samples of materials of the outside surfaces of the ISS Russian segment modules, which are exposed within the cassettes.

Physical Science and Materials Research Facilities

Super-High temperature Synthesis in space (SHS) [Roscosmos] This experiment is designed to develop a very interesting field of material science in space for fabrication and repair (welding, joining, cutting, coating, near-net-shape production, etc.) in microgravity and even on the moon and other planets. Russian scientists have a very good collaboration in this field of investigation on the ISS with other partners (Europe, Japan, Canada). This process is a combination of several gravity-affected physical and chemical processes, operating at temperatures of synthesis up to 3000 K.

View of SHS Process

Bar and Expert Experiments [Roscosmos] use a unique set of instruments for temperature cartography, ultrasonic probing, and pyro-endoscopic analysis of potentially dangerous places on board the ISS. Zones of possible formation of condensation have been revealed, and potential corrosion damage has been evaluated.

Physical Science and Materials Research Facilities

Earth and Space Science

The presence of the ISS in low-Earth orbit provides a unique vantage point for collecting Earth and Space Science data. From an average altitude of about 400 km, details in such features as glaciers, agricultural fields, cities, and coral reefs taken from the ISS can be layered with other sources of data, such as orbiting satellites, to compile the most comprehensive information available.

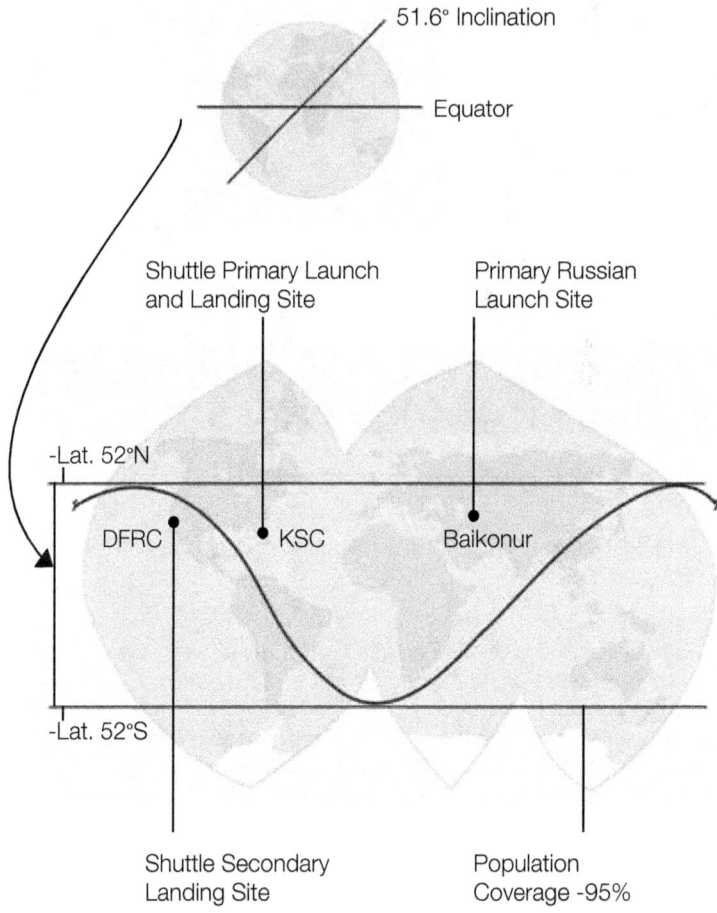

Planned ISS External Payload Attachment Locations

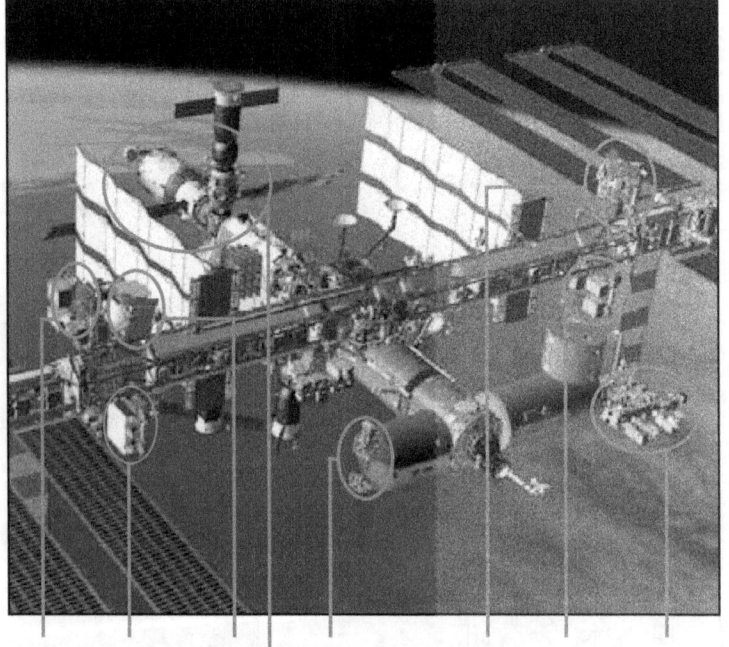

ELC-2   ELC-4   AMS | Columbus-EPF   ELC-3   ELC-1   JEM-EF

External Universal Workstations (9) on the Russian Service Module

Earth and Space Science Facilities

*External Earth and Space Science hardware platforms are located at various places along the outside of the ISS. Locations include the **Columbus External Payload Facility (CEPF)**, Russian Service Module, Japanese Experiment Module Exposed Facility **(JEM-EF)**, four **EXPRESS Logistic Carriers (ELC)** and the **Alpha Magnetic Spectrometer (AMS)**.*

Earth and Space Science Facilities

Columbus-External Payload Facility (Columbus-EPF) [ESA] provides four powered external attachment site locations for scientific payloads or facilities, and is being used by ESA and NASA. The first two European payloads on Columbus-EPF are major multi-user facilities in themselves. **EuTEF** (European Technology Exposure Facility) is a set of nine different instruments and samples to support multidisciplinary studies of the ISS external environment, from radiation and space environment characterization to organic and technology materials exposure. **Solar** (Sun Monitoring on the External Payload Facility) is a triple spectrometer observatory that is currently measuring solar spectral irradiance. Knowledge of the solar energy irradiance into the Earth's atmosphere and its variations is of great importance for atmospheric modeling, atmospheric chemistry, and climatology.

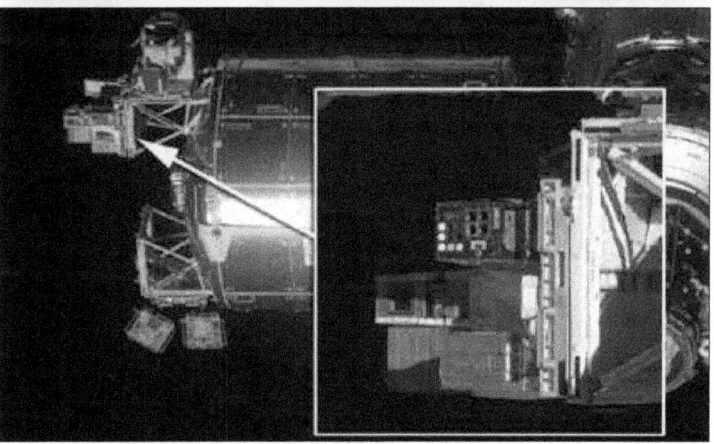

*Two external facilities, **EuTEF** and **Solar**, provide sites for a variety of external material science and solar research experiments. In the future the ACES payload with two high precision atomic clocks and the **Atmosphere Space Interaction Monitor (ASIM)** will be deployed on CEPF.*

Cosmic Ray Detectors and Ionosphere Probes [Roscosmos] are important for studies of cosmic rays and the low-Earth orbit environment. **Platan** is an external detector for cosmic rays, **BTN** is an external detector measuring neutron flux, and **Vsplesk** is an external detector for gammarays and high energy charged particles. Two packages, **Impulse** and **Obstanovka**, include ionosphere probes and pulsed plasma source (**IPI-100**) for making measurements of the ionosphere parameters and plasma-wave characteristics and are planned for launch and mounting outside ISS in the future.

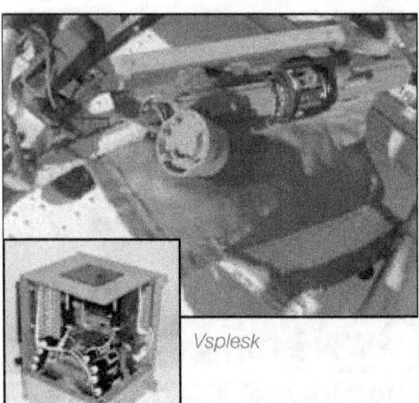

Vsplesk

Platan

The Global Transmission Services (GTS) Experiment is continuously operating within an ESA/Russian cooperation on the Russian segment of the ISS, and is testing the receiving conditions of a time and data signal for dedicated receivers on the ground. The time signal has special coding to allow the receiver to determine the local time anywhere on the Earth. The main objectives of the experiment are to verify the performance and accuracy of a time signal transmitted to the Earth's surface; the signal quality and data rates achieved on the ground; and measurement of disturbing effects such as Doppler shifts, multi-path reflections, shadowing, and elevation impacts.

BTN

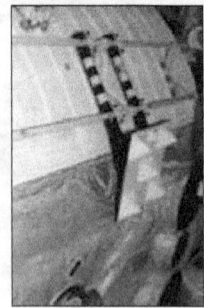

GTS Antenna

Earth Resources Sensing and Geophysics Instruments [Roscosmos] are used in studies of geophysics, natural resources, and ecology. **Fialka** is an ultraviolet imager and spectrometer used to study radiation emitted by reactions of propulsion system exhaust products from ISS, Progress, and Soyuz vehicles with atomic oxygen. It is also used to study the spatial distribution and emission spectra of atmospheric phenomena such as airglow. **Rusalka** is a micro spectrometer for collecting detailed information on observed spectral radiance in the near IR waveband for measurement of greenhouse gas concentrations in the Earth atmosphere.

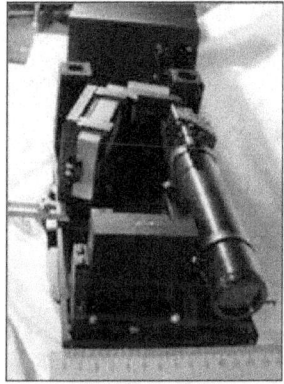

Rusalka

Fialka

Expedite the Processing of Experiments to the Space Station (EXPRESS) Logistics Carrier (ELC) [NASA] is designed to support external payloads mounted to the ISS trusses, as well as store external spares (called Orbital Replacement Units) needed over the life of ISS. Four ELCs will be delivered as part of the final assembly missions. Three ELCs are attached to the starboard truss 3 (S3) and one ELC is attached to the port truss 3 (P3). By attaching at the S3/P3 sites, a variety of views such as zenith (deep space) or nadir (Earthward) direction with a combination of ram (forward) or wake (aft) pointing allows for many possible viewing opportunities.

SEDA-AP   ICS   MAXI

JEM Exposed Facility (JEM-EF) [JAXA] is an unpressurized pallet structure attached to the Japanese Experiment Module (JEM), *Kibo*. This external platform will be used for research in areas such as communications, space science, engineering, materials processing, and Earth observation. The **ICS** (Inter-Orbit Communication System) is used to downlink data to Earth. The first JAXA experiments for the JEM-EF are **SEDA-AP** (Space Environment Data Acquisition equipment-Attached Payload, which measures the space environment around the ISS), **MAXI** (Monitor of All-sky X-ray Image, an instrument to monitor the X-ray sources in space), and **SMILES** (Superconducting Submillimeter-wave Limb-emission Sounder, which enables global observation of trace gases in the stratosphere), which will be located next to SEDA-AP.

Earth and Space Science Facilities

Window Observational Research Facility (WORF) [NASA] provides a facility for Earth science research using the *Destiny* optical-quality science window on the ISS. WORF provides structural hardware, avionics, thermal conditioning, and optical quality protection to support a wide variety of remote sensing instruments operating in the shirtsleeve environment of the pressurized ISS laboratory.

*Destiny features an Earth observation window with the highest quality optics ever flown on a human occupied spacecraft. The first remote sensing instrument to be used in **WORF**, **AgCam** (Agricultural Camera) is a infrared camera that will take frequent images of growing crops to help farmers manage their lands.*

Diatomia [Roscosmos] is an investigation aimed at the detection and study of ocean bioproductivity. Experiment "Seiner" is targeted on monitoring of ocean fish-rich areas, and on communication with fishing boats.

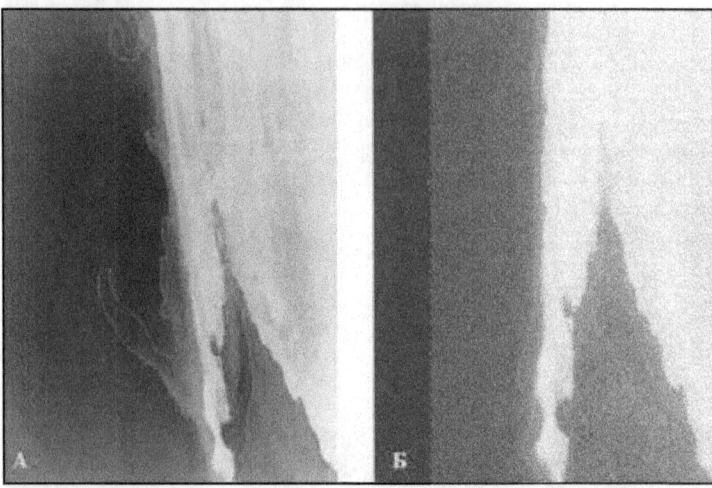

Geographical location and shape of high-productive water area of Canary upwelling in intensification in 2008 (A) and relaxation in 2005 (Б) periods.

The Alpha Magnetic Spectrometer (AMS-02) [NASA] is a state-of-the-art particle physics detector constructed, tested and operated by an international team composed of 60 institutes from 16 countries and organized under United States Department of Energy (DOE) sponsorship. The AMS-02 will use the unique environment of space to advance knowledge of the universe and lead to the understanding of the universe's origin by searching for antimatter, dark matter and measuring cosmic rays. As the first magnetic spectrometer in space, AMS-02 will collect information from cosmic sources emanating from stars and galaxies millions of light years beyond the Milky Way.

ISS Control Centers

CSA-Payloads Telescience Operations
Center (PTOC), St. Hubert, Quebec, Canada

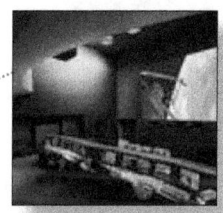

Canadian Space Agency Mission
Control Center (CSA-MCC),
Longueuil, Quebec, Canada

NASA - Payload Operations and
Integration Center (POIC), Huntsville, AL

NASA - Mission Control
Center (MCC), Houston, TX

ESA ATV - Control Center
Toulouse, France

*ESA-European User Support
Operations Centers:
CADMOS, Toulouse, France
MARS, Naples Italy
MUSC, Cologne, Germany
B-USOC, Brussels, Belgium
E-USOC, Madrid, Spain
N-USOC, Trondheim, Norway
DAMEC, Odense, Denmark
BIOTESC, Zurich, Switzerland
ERASMUS, Noordwijk, The Netherlands*

*ESA - Columbus Control Center
(Col-CC), Oberpfaffenhofen, Germany*

*HTV Control Center (HTVCC),
Tsukuba-shi, Ibaraki, Japan*

*Japan Experiment Module Mission Control
(JEMMC), Tsukuba-shi, Ibaraki, Japan*

*Roscosmos - Flight Control Center
(TsUP), Korolyov, Russia*

*Roscosmos - Transport Vehicle
Control Room, Korolyov, Russia*

To Learn More

Space Station Science
http://www.nasa.gov/mission_pages/station/science/

Facilities
http://www.nasa.gov/mission_pages/station/science/experiments/Discipline.html

ISS Interactive Reference Guide
http://www.nasa.gov/externalflash/ISSRG/index.htm

CSA - Canada
www.asc-csa.gc.ca/eng/iss/default.com

ESA - Europe
http://www.esa.int/esaHS/iss.html

JAXA - Japan
http://iss.jaxa.jp/en/

Roscosmos - Russia
http://knts.rsa.ru
http://www.energia.ru/english/index.html

Acronyms

ABRS	Advanced Biological Research System
ADUM	Advanced Diagnostic Ultrasound in Microgravity
AgCam	Agricultural Camera
ALTEA	Anomalous Long Term Effects in Astronaut's Central Nervous System
AMS	Alpha Magnetic Spectrometer
AQH	Aquatic Habitat
ARED	Advanced Resistive Exercise Device
ARIS-POP	Active Rack Isolation System-Payload On Orbit Processor
BioLab	Biological Experiment Laboratory
BISE	Bodies In the Space Environment
BSTC	Biotechnology Specimen Temperature Controller
CBEF	Cell Biology Experiment Facility
CEO-IPY	Crew Earth Observations-International Polar Year
CEPF	Columbus External Payload Facility

CEVIS	Cycle Ergometer with Vibration Isolation System
CGBA	Commercial Generic Bioprocessing Apparatus
CIR	Combustion Integrated Rack
CKK/SKK	Replaceable Cassette-Container
COL-EPF	Columbus External Payload Facility
COLBERT	Combined Operational Load Bearing External Resistive Exercise Treadmill
CUCU	COTS UHF Communications Unit
DECLIC	Device for the study of Critical Liquids and Crystallization
EDR	European Drawer Rack
ELC	EXPRESS Logistics Carrier
ELITE-S2	ELaboratore Immagini Televisive-Space 2
EMCS	European Modular Cultivation System
eOSTEO	Bone Culture System
EPM	European Physiology Module
EXPRESS	Expedite the Processing of Experiments to the Space Station
EuTEF	European Technology Exposure Facility
EVARM	Study of Radiation Doses Experienced by Astronauts in EVA
FIR	Fluids Integrated Rack
FSL	Fluid Science Laboratory
GHF	Gradient Heating Furnace
GLACIER	General Laboratory Active Cryogenic ISS Equipment Refrigerator
GSM	Gas Supply Module
GTS	Global Transmission Services
HPA	Hand Posture Analyzer
HRF	Human Research Facility
ICS	Inter-Orbit Communication System
ISIS Dwr	International Subrack Interface Standard
ISPR	International Standard Payload Rack
ITS P3	ISS Truss Site Port 3
ITS S3	ISS Truss Site Starboard 3
JEM-EF	Japanese Experiment Module-Exposed Facility
LADA	Greenhouse
LOCAD-PTS	Lab-on-a-Chip Application Development-Portable Test System
MAMS	Microgravity Acceleration Measurement System
MARES	Muscle Atrophy Research Exercise System
MAXI	Monitor of All-sky X-ray Image
MDS	Mice Drawer System
MELFI	Minus Eighty-Degree Laboratory Freezer for ISS
MERLIN	Microgravity Experiment Research Locker/Incubator
MISSE	Materials International Space Station Experiment
MSG	Microgravity Science Glovebox
MSL	Materials Science Laboratory
MSPR	Multipurpose Small Payload Rack
MSRR	Materials Science Research Rack
PCDF	Protein Crystallization Diagnostics Facility
PEMS	Percutaneous Electrical Muscle Stimulator
PFS	Pulmonary Function System
PGB	Portable Glove Box
PMDIS	Perceptual Motor Deficits In Space
Portable-PFS	Portable Pulmonary Function System
SAMS-II-ICU	Space Acceleration Measurement System-Interim Control Unit
SEDA-AP	Space Environment Data Acquisition equipment-Attached Payload
SLAMMD	Space Linear Acceleration Mass Measurement Device
SMILES	Superconducting Submillimeter-wave Limb-emission Sounder
SpaceDRUMS	Space Dynamically Responding Ultrasonic Matrix System
VCAM	Vehicle Cabin Atmosphere Module
WORF	Window Observational Research Facility

www.ingramcontent.com/pod-product-compliance
Lightning Source LLC
Chambersburg PA
CBHW061518180526

45171CB00001B/237